Foreword

Everyone is a node

Zizenwallet.eth / haztinzvgyge July 12, 2018

The most famous spirit of the blockchain is "decentralization", which can neither be modified nor stolen, and is anonymized. Blockchain is suitable for financial applications in different scenarios. Currently, Bitcoin has the highest market value and is the most well-known application. The power of money must be admirable. Technology and Finance have always been the mutual kinetic energy of the growth of the economy. In this period, the combination of both -which is called FINTECH- will disturb many things in the world.

The new technology that changes the financial tradition is gradually having the same cruel ecological performance because of the intervention of traditional finance, and is much more faster. Some people joked that the cryptocurrency and blockchain market are decentralized capitalism. In fact, capitalism has always been decentralized. Our social consensus is not able to conclude the convergence of blockchains, nor the pursuit of decentralization. But the consensus must begin to assemble, and money is the basic that drives the decentralized consensus assembly. Money itself is faith, and money reflects the value of communication. We have no power to determine which is good, which is bad. We can only choose in our own way, choose to love, choose to hate, or choose to ignore them. The design of the node mechanism is an application of blockchain technology. It is used in the different way as the mining machine to achieve the same goal- maintenance of the blockchain. This is the establishment of a consensus. The purpose of the node is to assist the blockchain to reach consensus and confirmation, and to benefit from it. Each node must maintain community consensus, maintain the best interests of the community, and then achieve their own interests. At the time of the arrival of this new technology, the human community is carrying out the largest wave of migration in history. Everyone must uphold the correct path in their hearts. Confirm and interact with other people's information. Reach consensus and adjust their direction.

每個人都是節點

區塊鏈最著名的精神是"去中心化"，而其不可篡改性，與附加的匿名性、匿蹤性等等，適合用於不同場景的金融應用，目前市值最高的比特幣，為最知名的第一個應用。

金錢這股力量，是必須要敬畏的。

科技與金融，一直是經濟體成長的互為驅使的動能，而在這個時期兩者互為一體，更是將擾動這個世界上的許多事情。

我們可以看到這改變金融傳統的新科技，也逐漸的因為傳統金融的介入，看到一樣殘酷的生態上演，更加快速。

有人戲稱加密貨幣、區塊鏈市場，是去中心化的資本主義，實際上資本主義一直都是去中心化的。

區塊鏈在這個時代的交匯，追求去中心化的意義，我們的社會共識暫時無法有所結論，但共識是必須開始集結的，金錢則是驅動去中心化共識集結的中心。

金錢本身即是信仰，金錢是反映了溝通上的價值。我們沒有權力去決定孰善孰惡。我們每個人都只能用自己的方式進行選擇，選擇愛、選擇恨、選擇無視。

節點機制的設計，是區塊鏈科技上的一種應用，與礦機演算用不同的方式，達到同樣維護區塊鏈的目的。這是一種共識的建立。

節點的目的,是協助區塊鏈達成共識以及確認,而從中取得利益。每個節點都必須維持社群共識,維持社群最大利益,進而達到自己的利益。

這個新科技來臨的時節,人類社群正進行一種史上最大的數位遷徙潮,每一個人都必須秉持自己心中正道,與其他人的資訊交互確認,彼此取得共識、調整方向。

Table of Contents

1 Introduction

1.1. What is a Masternode?

1.1.1. The three types of Blockchain

1.1.2. Full nodes and their relationship to Masternodes

1.1.3 Block Producers and their differences with Miners and Masternodes

ENUVIP (enuviptaiwan)

Enu One Family (account name: enuonefamily)

1.2. Selection of Masternodes in the Blockchain

2 Importance of Masternodes

2.1 Increasing the privacy of transactions

2.2 InstaSend

2.3 PrivateSend

2.4 Decentralized Governance

3 What is required to host a Masternode?

4 How Masternodes make money

5 Choosing VPS services for Masternodes

5.1 Advantages

5.1.1 Reliability

5.1.2 Control

5.1.3 Scalability

5.1.4 Cost savings

5.1.5 Support

5.2 Disadvantages

5.2.1 Oversold resources

5.2.2 Expenses

5.3 Choosing the best VPS host for a Masternode

6 Maximizing revenues from Masternodes

6.1 Coin selection for a Masternode

6.2 Top coins for Masternode investments

- 6.2.1 Dash
- 6.2.2 ALQO
- 6.2.3 MUE (Monetary Unit)
- 7 Practical set up of a Masternode
- 7.1 SmartCash Masternode
- 8 Trends in Masternodes
- 8.1 Investing to own a Masternode
- 8.2 Speculation
- 8.3 Multiplicity of Masternodes
- 8.4 Adoption by other cryptocurrencies
- 8.5 Rising cost of ownership
- 8.6 High ROI Masternodes scam
- 8.7 Artificial scarcity
- 8.8 Weak coins

Conclusion

Acknowledgement

1 Introduction

1.1. What is a Masternode?

A Masternode can be simply explained as a computer wallet, in particular, a dash wallet. The computer is used to keep a copy of the Blockchain in real time. The Blockchain is the heart of all cryptocurrencies, it is the digital ledger that is incorruptible and is used to store digital transactions. It is a distributed database that is shared across a network of computers and is continually updated. Information on the Blockchain is incorruptible because there is no centralized location that can be compromised to modify all the data at a go. Blockchain also has no single point of failure. There is too much replication such that it is almost impossible to take down the network it relies on.

To best understand what a Masternode is, it is good to understand the concept of a node in Blockchain. A node is a computer connected to the Blockchain network that only validates and relays transactions. All nodes have a continually updated copy of the Blockchain. Together, many nodes form a powerful network that facilitates the services offered by Blockchain. Nodes join and leave the network voluntarily. Since the tasks they carry out are resource-intensive, they are incentivized by the Blockchain. Whenever a node is in the Blockchain network, it stands a chance of winning some cryptocurrencies if it successfully facilitates a transaction. Therefore, nodes compete to get the chance to solve the complex calculations that facilitate transactions through the Blockchain.

1.1.1. The three types of Blockchain

With a better understanding of nodes, we can now go deeper into what a Masternode really is. The clearer definition lies deeper inside the Blockchain technology. It might seem that Blockchain is a single type of technology that relies on a giant peer-to-peer network while in reality, it is actually three types. These are Proof of Work, Proof of Stake and a combination of these two. These three types of Blockchain need to have an incorruptible way of processing their transactions in order for the whole technology to function well.

Proof of Work is the rudimentary type of Blockchain technology. However, it is still advanced when compared to traditional methods of payment that are still in use today. POW underpins a trustless distributed consensus system of transacting. This is a system where neither of the transacting parties has to trust a third party to complete a transaction. In fiat currencies, a third party is required such as a bank, PayPal, and Payoneer among others to complete a transaction between two parties. POW removes the third party by ensuring that a transaction can be completed by the transacting parties themselves. POW is what ensures this can happen.

This is the technology that has been discussed above where nodes on the Blockchain network compete to solve complex puzzles in order to be rewarded with Bitcoin or another specific digital currency. Miners underpin the POW Blockchain and it cannot function without them. They do all the hard work of processing and validating blocks of transactions by solving complex equations. Mining also ensures that new coins are created but there are set limits on the number of coins that can be created per each type of digital currency. However, the POW Blockchain has been energy-intensive with miners (people that run nodes on the Blockchain) having to spend a lot of money on acquiring hardware capable of mining enough Bitcoin to barely make some profit. The POW Blockchain has also been said to be rudimentary and therefore there is a shift from POW to POS. Ethereum, in particular, has said that it will hard fork so as to move from POW to POS. Many other cryptocurrencies are expected to do the same. The cost of POW is only going up. Miners are using more resources in terms of processing power and energy. It is said that the current requirement to process just one Bitcoin transaction is equivalent to the electrical energy consumed by one and a half households in America per day. These costs are being paid for using fiat currencies which are in turn costing too much on miners. At the current rate of growth of energy demand to process a single Bitcoin transaction, it is said that by 2020, a single transaction will take up as much electrical energy as Denmark consumes daily. This will be unsustainable thus the rising interest in the POS Blockchain technology.

Proof of Stake (POS) is the type of Blockchain that is supported by Masternodes. POS is also a way of validating transactions but in a totally different way than POW. The Masternode is itself a computer that performs several functions on Blockchain. These functions might include; adding privacy, effecting instantaneous transactions or even data storage. A Masternode connects to the Blockchain network and waits in a queue to be selected to perform the available tasks. Selection of a Masternode is done in a deterministic way using an algorithm. This is unlike the POW algorithm where the chance of getting selected is non-deterministic. Many POW coins are adapting to Masternodes. They are using Masternodes alongside the nodes used for mining or in simpler terms, the nodes that solve complex equations to validate a cryptocurrency transaction.

1.1.2. Full nodes and their relationship to Masternodes

To add context to what a master node is, we can take a look at a special node in the Blockchain is the called a full node. A full node normally keeps an updated copy of the

Blockchain of its particular coin. It then spreads the updated copy to all the nodes run by miners. For instance, a Bitcoin miner will receive a copy of the Bitcoin Blockchain from a Bitcoin full node. These full nodes are only responsible for spreading the updated copies of transactional data. Without full nodes, it would be hard to achieve the decentralized nature of the Blockchain technology. A central server would have to continually update all nodes in the network with new data and this would be extremely difficult. Full nodes make it possible for the data to be distributed to all nodes is a reliably efficient way. They also ensure consistency in the Blockchain. They ensure this by:

- Validation and relay of transactions and blocks
- Synchronizing nodes in the network that were offline
- Filtering transactions and blocks for other nodes

Full nodes are trustless and they check all blocks against a coin's core consensus. Lightweight nodes only download block headers and use third-party websites to validate them. Full nodes are also the backbone of security in the Blockchain. They audit the work done by miners and the information being transmitted by other users on the network. The biggest challenge with full nodes is that there are no defined incentive structures and most people that run them do so for the purpose of helping a network. The lack of incentives has seen many people withdraw from offering their high-end machines as full nodes. In 2016, there were approximately 200,000 Bitcoin full nodes but by early 2017, the number sharply declined to just about 6,000. The withdrawal of Bitcoin full nodes has posed several threats to the coin and is affecting its value. There are risks that without full nodes, there is a looming centralization of the coin. This is because lightweight nodes are meant to follow nodes that have most mining power. Since the entry costs of mining have already gone high, mining pools made up of expensive computing hardware are centralizing Bitcoin. Centralization is going to expose Bitcoin to vulnerabilities by introducing weak spots in its network that hackers can strategically attack. Nevertheless, the importance of full nodes in terms of the roles that they play in the Blockchain can be appreciated with a view of what their absence causes.

A Masternode can do all the functions that a full node does and even more. Masternodes play more roles in a Blockchain network than full nodes. They are of more importance than full nodes in any coin's network. Master nodes are at times referred to as full nodes that can perform validations and offer services that miners cannot. Unlike full nodes that are not rewarded, Masternodes have an economic incentive to keep playing their roles. Masternodes are different from nodes and full nodes in that they require a collateral in order to operate. Therefore, each Masternode has to have a stake in the coin that it serves. This way, when the value of the coin goes up, the Masternode gains more value. The bond is also used to ensure that Masternodes carry out their functions accurately. Whenever a Masternode wrongly validates a block, its bond is slashed. Lastly, the bond reduces monopoly of control. It is often a steep amount to part with and this reduces the probability of one party controlling most of the Masternodes in the Blockchain. There are different incentive structures for Masternodes that we shall discuss later.

1.1.3 Block Producers and their differences with Miners and Masternodes

Mining in cryptocurrency is normally defined as the process in which cryptocurrency transactions are added to the block chain. The Blockchain is a public ledger and a block is equated to a page in the ledger. Mining entails the compilation of all recent transactions into a block as well as the computation of difficult puzzles in order to validate the transactions. There can only be one correct answer to the complex puzzle and the first participant to get the right answer is given the privilege to create the next block in the open ledger. As an incentive, the participant is also invited to claim a reward in terms of coins and the transaction fees of the compiled block. Miners and block producers are key players in coins that use PoW. However, they are different players and play distinct roles.

Block producers came into play with EOS Blockchain technology. Their work is to validate blocks and provide key infrastructure for the Blockchain. There is a predetermined number of block producers in the coins that use them. This is unlike miners and Masternodes that tend to be unlimited in number due to the lack of entry barriers. Block producers earn their rewards in terms of EOS tokens and these are produced through inflation. For self-sufficiency, EOS normally has an inflation of 5% but 1% goes to the payment of block producers. The remaining 4% is normally set aside to be used by a proposal system. In contrast to this, miners earn block rewards and transaction fees while Masternodes are rewarded with a percentage of a block's revenues. Miners are rewarded individually while Masternodes are rewarded as a group.

To further understand the concept of block producers, a keen look at EOS is required. EOS came with features such as parallel processing, a set of rules in a constitution, a self-sufficiency mechanism of annual inflation and a decentralized OS. The decentralized OS is an important part of the EOS technology that allows developers to build different apps on it. This OS is normally hosted on data centers and they are the ones referred to as block producers. In addition to providing the infrastructure to host the OS, they validate and produce blocks. Block producers in EOS normally have to first announce their candidature to become one. There are eight criteria that are used to gauge BP candidates. These are:

- Public presence via a web URL or a social network handle,
- The candidate's ID information on Steem Blockchain
- The candidate's technical specs for the hardware resources they own
- The candidate's plan to scale up their hardware
- An outline of the benefits the community will get from the candidate
- Telegram and Test-net node names
- The candidate's roadmap in terms of finances, transparency and future direction
- The candidate's position on dividend sharing of inflation rewards

ENU

ENU is a coin that has cloned the EOS Blockchain technology. ENU started with one block producer before its voting system was completed. Ever since then, the ENU community votes for block producers. Block producers are listed on the ENU Block Producers Monitor (http://enu.6btc.com/) and their activities updated in real time. To become an ENU Block producer, one has to be well known by the community in order to get their votes. There are several Telegram channels and discussion platforms that one can join to introduce themselves to the community. ENU block producers are rewarded by ENU tokens and they can claim them after they have validated a given number of blocks. ENU currently has a total of 21 active block producers. There is a list of 66 that the community can browse through to determine the one they wish to vote for based on the 8 criteria we discussed above.

An integral part of ENU is UBI (Universal Basic Income) which is an idea of giving an equal share of resources to all humans. The project is based on the understanding that, just as everyone has the right to breathe air, they have a right to equitably enjoy the planet's resources. Enumivo has come up with a solution to achieve this by providing a decentralized app that can be used to ensure that humans can earn their equitable share of resources through a regular income provided using UBI tokens. To ensure scalability, speed and cost effectiveness, the app will use a verification system that is a clone of EOS. Through the equitable share of resources, the UBI app will settle the existing financial imbalance in many parts of the world. The app will also offer other incentives such as education. The UBI app is set for release in early 2019 with all unique users entitled to 20 UBI tokens per week for 2500 weeks.

We believe in ENU and its mission to ensure equal share of resources. We have seen real-life implementations of similar projects though by donors across Africa. Therefore, there is a lot of potential in the project both in the short and the long term. We are active producers on the project and we have gained credibility. Currently, we have the following:

ENUVIP (enuviptaiwan)

https://enuvip.com/

This is an ENUMIVO community based in Asia that views everyone as a VIP. The community has a talented tech team that ensures software and hardware readiness for server hosting. The community is made up of cryptocurrency investors that are passionate about sharing the Blockchain technology with new users. Its members are from Taiwan, China and Malaysia.

You can join the group and become a royal member through this link:

https://line.me/R/ti/g/l30rRRvOQY

Enu One Family (account name: enuonefamily)

This is another Enumivo community that acts like a family. The community is focused on the following:

- Building a world-class server infrastructure so as to become a scalable block producer
- Hosting offline research and development parties

1.2. Selection of Masternodes in the Blockchain

We have mentioned that nodes in a cryptocurrency Blockchain compete to participate in validating cryptocurrency transactions. On the other hand, Masternodes are selected in a deterministic way. Masternodes are first placed in a validator pool. This is the pool where any new Masternode will be placed and it is open to all. There is no priority scheme in joining the pool, a Masternode can join at any time.

2 Importance of Masternodes

Masternodes are of great importance to the Blockchain they operate in. They handle tasks that cannot be handled by other nodes. That is why some of the mandatory requirements for a Masternode is to always be available and always be accurate. Failure to meet some of these mandatory requirements leads to financial loss through the slashing of the bond in the Masternode. However, what are these crucial tasks that Masternodes perform? They are as follows:

2.1 Increasing the privacy of transactions

Masternodes are responsible for anonymizing the Blockchain. They add a layer of protection of the parties that undertook a transaction. This makes it hard for the parties to be traced. There had been increased cases of transacting parties being arrested after their transactions were deanonymized by law enforcement agencies. Coins that have adopted Masternodes have become highly anonymous. A good example of such a coin is PIVX. Through Masternodes, PIVX enables its users to transact with their coins separated from the transaction IDs. This results in a completely anonymous transaction, but one that PIVX can still verify. In order for the separation to occur, users have to convert their normal coins to zPIV tokens and the Masternodes will handle the separation of identities when a transaction is done using such tokens.

2.2 InstaSend

One of the greatest advantages of Masternodes is that they can provide instant transactions. This is something that other nodes cannot provide. In 2018, an average Bitcoin transaction will take 17 minutes (). A Dash transaction, on the other hand, takes just a few seconds. The big difference is because Dash has Masternodes while Bitcoin does not. This eventually leads to Bitcoin transactions taking long. Even though the confirmation in Dash takes a few minutes, the transaction is instantaneous and it shows in a receiver's wallet seconds after it is executed by the sender. A series of checks is normally put in place to ensure that the InstaSend mechanism cannot be exploited. When a sender executes a transaction, the Masternode will lock the funds such that the sender cannot double spend before a confirmation is given. The transaction will be added to a future block but the transaction is marked as successful since the Masternode has already locked the funds thus guaranteeing that they will be paid. In cryptocurrencies such as Bitcoin that do not have Masternodes, senders have to wait till their block is processed for the transaction to be completed. This is because nodes do not have the privileges that Masternodes have or use to effect instant transactions.

2.3 PrivateSend

This is a feature that has added the privacy of transactions. It is different from the mechanism that was previously discussed. PrivateSend can be compared to CoinJoin. Coinjoin is a method of anonymizing Bitcoin transactions by merging several transactions thus obfuscating the original ones. Therefore, several senders will have to jointly mix their coins and then do a joint transaction. A transaction in a cryptocurrency such as Bitcoin has two addresses an in and an out. The "in" address is the address from which the coins were sent from and the out address is the address to where the coins were sent to. All these addresses are stored on the Blockchain, a ledger that is publically accessible. Therefore, it is possible for a motivated entity, such as a law enforcement agency, to look into the Blockchain records and trace back the transacting parties through the in and out addresses. CoinJoin, however, makes this difficult by combining many inputs and outputs into a single transaction thus effectively ending the traceable digital trail.

Coins that have Masternodes do not have to rely on CoinJoin as the Masternodes handle this burden for users. Masternodes will take several transaction inputs and merge them before putting them on the Blockchain. A merge is done with at least 3 participants and once the Masternodes have coupled the transactions, they cannot be uncoupled. This, therefore, makes it hard for an entity to try and go through the Blockchain of a coin such as Dash to try and uniquely identify the addresses of transacting parties.

2.4 Decentralized Governance

Masternodes have the privilege of voting in their Blockchain. Voting is done when some decisions have to be made about some technological or financial developments. Each Masternode has one vote and this is used to vote for proposals. When many Masternodes vote in a proposal, it is adopted. When the majority of Masternodes vote against a proposal, it is rejected. Therefore, owners of Masternodes have a say in the developments of their cryptocurrency. This is part of the reason why they have a stake in the cryptocurrencies. It is just like in the real world where shareholders have voting rights. They can use these rights to influence the way a company operates. A common competition technique in the real world is whereby a competitor buys more than 50% of the shares in a company. The competitor is, therefore, able to control the development of the company in his or her favor. In the case of Masternodes, it is extremely hard for a similar thing to occur. This is because of the initial high purchase price of collateral to own a Masternode. It is therefore prohibitively expensive for a malicious party to own the majority of votes. There is also the aspect that the number of Masternodes keeps on changing.

3   What is required to host a Masternode?

In the previous chapters, it was mentioned that Masternodes are POS Blockchain. Therefore, they have to have some stake or ownership in the coin they serve. There are different lists of requirements depending on the coin that one wishes to invest in. The most important requirement is the number of coins that one requires to have as bond or stake in the coin. Masternodes generally require the owners to purchase lots of coins. However, these coins can still be extracted and sold if the owner so wishes. Without the required number of coins, it is impossible to own a Masternode. Due to this, whenever Masternodes are activated for a certain coin, the number of coins in circulation decreases and this pushes the price high.

The other important requirement is the computer and network setup. It is mandatory for a Masternode to be available at all times. The Masternode should at least have 1GB RAM, 1CPU core, and 20GB HDD space at all times. Even more important, the Masternode needs to be almost always online since it is a server being relied on by numerous hosts in a Blockchain. The Masternode must also have a static IP address. A static IP address means that the server's address does not change from time to time.

Next, one is required to have a local wallet and a hot wallet for the Masternode. There are different types of wallets available depending on the computer that one is using. Once a wallet has been installed and set up on the computer, a few things need to be done. One has to activate coin control or Masternode features based on the wallet that he or she is suing. The wallet to be used by the Masternode needs to be credited with the exact amount required for the specific coin that one is working with. The amount should not be more or less than the required collateral. For instance, if coin X requires 10000 coins to run a Masternode, the wallet to be used by the Masternode needs to be credited with exactly 10,000 X coins. Once these coins are received, they have to be locked in the wallet according to the guidelines of owning a Masternode. The lock prevents these coins from being spent. If by any chance these coins are taken out of the wallet, this will lead to the Masternode being kicked out due to the lack of the required collateral. This is why these coins have to be kept in a cold storage.

For convenience purposes, most owners of Masternodes prefer working with Virtual Private Servers instead of local computers since VPSs can offer more reliability. Some VPS vendors are economical in their charges and offer 24/7 support as well as 99.9% guaranteed uptime. This is a good deal for a Masternode. With the VPS set up and the cold storage wallet with the required collateral, one is ready to run a Masternode.

4 How Masternodes make money

There are different ways in which Masternodes make money. However, we will refer to how to the first coin to host Masternodes set up the payment structure for Masternodes. It is given that a Masternode will be paid at the end of the day for the tasks that it assists such as PrivateSend and InstaSend. In Dash, the Masternodes were set to be paid 45% of the block reward. The block is like a page on the Blockchain distributed ledger that records new transactions. Therefore, once the block is successfully filled by a number of transactions, the reward for these transactions are shared with miners and Masternodes.

In Dash, after every 16,616 blocks, there is a creation of a superblock. A superblock contains payouts for budget proposal winners as voted by the Masternodes. The selection of Masternodes to be paid per block is done randomly but in a deterministic way. This means that, even though all the Masternodes do not get paid at the same time, they will eventually get paid. A Masternode is selected for payment when it reaches the top 10% of the unpaid list of Masternodes. Once it receives a payment, it is sent back to the bottom of the list. The selection in Dash was done each 2.6 minutes though the duration varies with the number of Masternodes. The more the number of Masternodes, the more the duration between payments since all the Masternodes have to be paid. Coins that have Masternodes give the estimated daily, weekly, monthly and annual earnings since the algorithm used is deterministic and even though all Masternodes do not get paid simultaneously, the earnings in the long term will be almost the same.

There are instances where a Masternode may stop getting payments. This normally arises from violations of some rules. There are two strict rules that Masternodes must always adhere. These are; having the required collateral and offering reliable services. Therefore, if the Masternodes collateral is spent, it is kicked out of the network and payments to it are ceased. In the same vein, if a collateral becomes unavailable for more than an hour, it is deemed unreliable and kicked out of the network. These two rules ensure that Masternodes are always motivated to retain the collateral balances and to provide reliable services.

5 Choosing VPS services for Masternodes -

The best option when setting up a Masternode is to use a Virtual Private Server instead of a local computer or a shared server. This is VPSs run in a more controlled and reliable environment and the vendor takes responsibility for very many burdens such as ensuring 99% uptime and high-speed internet access. The following is a more detailed explanation of the pros and cons of using a VPS:

5.1 Advantages

5.1.1 Reliability

Virtual Private Servers are more reliable than local servers or shared servers. When one runs a Masternode from a local server, he or she must meet all the maintenance tasks of the server. Additionally, they must ensure that the Masternode has a high-speed internet connection and is in a backed-up power supply. This makes the initial costs of running a Masternode on a local server high without even the assurance that the server will have the required uptime. Shared servers, on the other hand, are unreliable due to the shared environment and the issue of a bad user carrying out actions that may cause the whole server to crash. VPSs, therefore, come in as isolated computing environments that vendors ensure will have a reliably high uptime and will always be connected to fast internet.

5.1.2 Control

The context of control comes in when comparing a VPS to a shared server. A VPS user is able to tweak the server to meet his or her needs. They can install the software they want, open or close some ports and remove software that they do not require. This makes it ideal for setting up a Masternode as one can optimize it and add security features anytime they want.

5.1.3 Scalability

Sometimes the requests to a Masternode may increase or the programs running on the server may start to overwhelm its capacity. However, since VPSs are merely instances running on a much larger server, they are not limited to the resources that one selected when doing the initial purchase. One can simply choose to increase the computational resources that the VPS has in order to make it run comfortably at the workload it is exposed to.

5.1.4 Cost savings

One of the biggest burdens of running a Masternode on a local server is maintaining the uptime. The server has to be cooled, needs to have uninterrupted power and has to have a high-speed internet connection. All these are costly and the owner has to meet them alongside the high initial cost of ownership. However, VPSs are rented and thus there are no initial cost overheads or other recurring charges for maintenance.

5.1.5 Support

When running a local server, the owner has to take responsibility for any errors that the machine may have. The lack of technical skills may limit one from being able to rectify these errors. With a VPS, the vendor takes responsibility for these errors. Additionally, most vendors will assure of 24/7 support whenever any issues arise. The support team will be able to handle any errors that one runs into with their server.

5.2 Disadvantages

5.2.1 Oversold resources

Some greedy vendors oversell the resources that their physical servers have. This is because they rely on a common belief that a user will not consume all the resources that they have paid for. The vendor may, therefore, have oversold the resources and in an unfortunate scenario, all the users use the resources they have paid for to the maximum. This may lead to problems especially for those running Masternodes as the server will not have sufficient resources to maintain its uptime. It might even get kicked out as a Masternode.

5.2.2 Expenses

VPSs are more expensive than shared servers. This is because they run isolated on the physical server and demand more resources than shared servers. It may, therefore, be cheaper to run a shared server that is capable of handling the requirements of a Masternode.

5.3 Choosing the best VPS host for a Masternode:

There are a number of companies offering VPS services today. These include Microsoft, Google, and Amazon among other companies. These three companies are big and are trusted by many people. However, how does one choose the perfect VPS? The checklist for selecting a VPS for the purpose of running a Masternode has the following items:

- A static IP address
- High uptime (greater than or equal to 99%)
- Configurability

Most companies will offer these and ultimately, it may run down to individual taste to choose the best VPS hosting provider. However, the following are considerations that one should put in place to help with the ultimate choice.

- Price range – charges for VPSs depend on the resources and services that one wants. Most hosting providers will have monthly charges and these run from $2.50 to $25. The more the hardware resources and internet bandwidth that one wants, the more the cost.
- Purpose – some web hosting companies give the option to customers to buy plans that allow them to include more than one domain name. A Masternode can as well be used to run a website or a blog on the side since it is a server that has a static IP address. Therefore, if one has other interests such as running different services on the Masternode server, they should consider buying a VPS with lots of resources. However, it is normally not advised to run anything else on a VPS that runs a Masternode for security purposes. If for instance one runs a website and the

website is compromised, hackers will have direct access to the Masternode through the breached website and they can steal all the coins in the Masternode. Therefore, all resources bought should only be dedicated to running the Masternode.

6 Maximizing revenues from Masternodes

6.1 Coin selection for a Masternode

Masternode operators earn differently depending on several factors. One of the most important factors is the coin that one has chosen. The revenue model of Masternodes is such that if a coin gains value, all the Masternodes earn more. Similarly, coins that do poorly lead to Masternodes making little or no profits. Therefore, to maximize revenues, one has to select a coin that is doing well compared to others and has a high Return on Investment. The coin should also be expected to appreciate in value over time. This is because the number of Masternodes will keep increasing and thus the shared revenue between all Masternodes will reduce. This is because earnings are shared among all Masternodes and the more the number of Masternodes, the lower the percentage of revenue shared with each. Since there are no entry barriers to the ownership of a Masternode, the most important thing is to look for coins that are expected to gain value quickly. This way, even if the revenue-share percentage is low, the actual amount shared is high due to the coin gaining in value.

To find out the coins that give the most attractive rewards, there are a number of comparison sites that have been set up with these stats. One of these is Masternodes Pro and its screenshots are shown below:

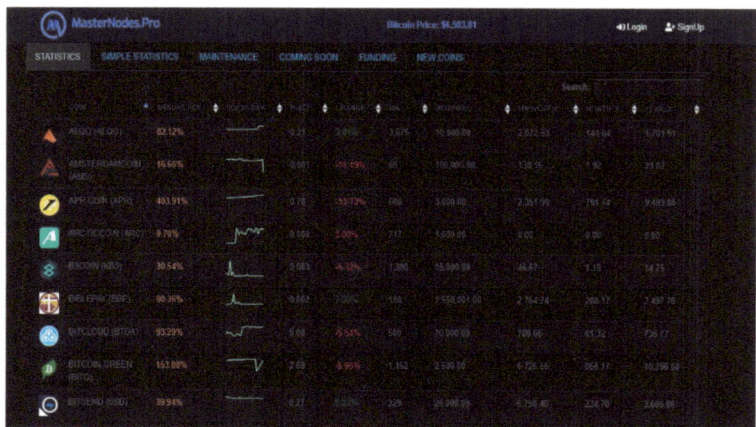

Figure 1: Full stats per coin (https://masternodes.pro/statistics)

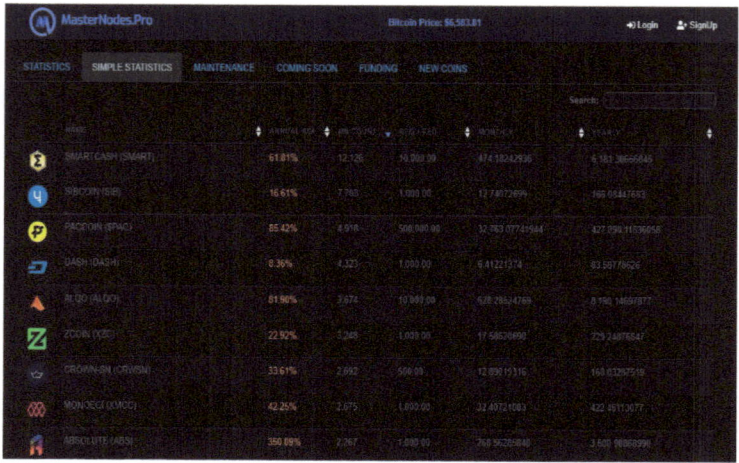

Figure 2: Simplified stats per coin (https://masternodes.pro/simple_statistics)

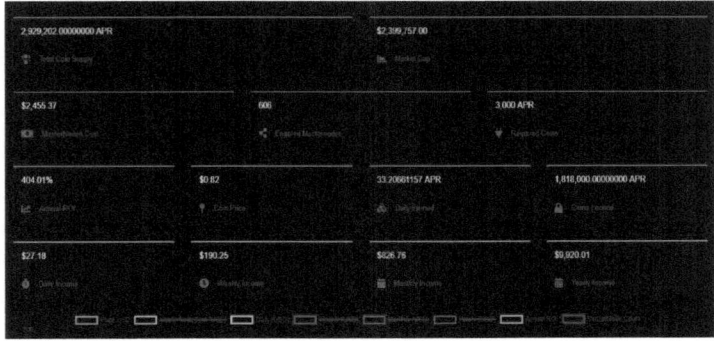

According to the screenshots provided, one can easily see that there are coins with promising annual ROIs. For instance, the figure one shows that the expected annual ROI of APR Coin is 403.91%. According to the graph given, it can be seen that the coin has been on a steady increase. This means that the coin is ideal for revenue generation using a Masternode. The cost of APR Coin as per the table is only $0.78. The minimum number of coins to have is 3000. This means that with $2340, one can have a Masternode for APR Coin. Upon clicking on the coin to get more details, the website shows the following (note that the price of the coin was still shooting up thus had moved from $0.78 to $0.82):

Figure 3: APR stats

According to this screenshot, APR is projected to give a daily return of $27.18, a weekly return of $190.5 and annual return of $9920. The ROI for this coin is 404% because when the annual income is divided by the initial cost of the Masternode, the outcome is 4.04 which translates to 404%. According to APR Coin, Masternodes receive 70% of the block rewards and this might be the reason why the returns are so lucrative.

These projections can be deceptive at times especially for relatively new coins. It is, therefore, necessary for one to do deeper research on why the coin will keep on gaining value. This is to help prevent an embarrassing outcome such as the one witnessed where Bitcoin soared to the high of $20,000 before plummeting down and causing people to lose entire investments made on the coin. A quick inspection of APR shows that it is focused on creating loyalty programs for companies. The parent company plans to offer APIs that businesses will use for their loyalty programs. However, the company will offer these APIs free of charge. Customers will also have the ability to exchange the loyalty coins that they have accumulated. APR will allow the exchange of APIs across brands. Therefore, at the moment the company has not yet started its actual operations. When APR starts operations, the value of the coin could appreciate and thus the projected revenues could still go up.

Another determinant of revenues a Masternode will get at the end of 12 months is the number of Masternodes. Unlike miners that get their individual revenues when they solve complex equations, all Masternodes in a network get revenues for each block processed. Therefore, the more the number of Masternodes, the less the profits. However, APR is not crowded according to the screenshot. The coin only has 606 Masternodes in its network. APR does not have any miners since it relies solely on POS Blockchain. Therefore, almost all the work regarding transaction processing is done by the Masternodes. As the coin grows, it will need more Masternodes. The initial cost of owning a single APR Masternode is not high. This means that many people could still setup Masternodes for the coin. This will, therefore, affect the annual revenues.

6.2 Top coins for Masternode investments

A detailed analysis of the cryptocurrencies that are offering Masternodes in their networks is required before one can choose to invest in one. The challenge with Masternodes is that the coins give their own estimates of the ROI but this is at times given to attract new Masternodes. There are too many weak coins that are offering attractive deals on their networks for Masternodes. In the end, one could end up getting burned if the values of these coins dip instead of rising. Therefore, one has to dig deep into the coin and understand its future prospective and its probability of growth. When the value of the coin stagnates or the coin fails to pick up, it is the Masternode investor that will feel the loss. Alongside this, one has to check just how complex it is to set up a Masternode. There are some coins with complex procedures for setting up and running a Masternode and this could prove to be problematic to users. The following are some coins which have the possibility of giving good returns for Masternode owners:

6.2.1 Dash

This coin is credited as the father of Masternodes since it was the first coin to have them. It is a fork of Bitcoin and is trusted by many people. The current ROI for Dash is at 8% and one is required to own 1000 coins to run one. At the cost of $238 per coin, one is required to part with $238,000 to own just one Masternode. The daily income is at $55 and the annual income is projected to be $20,000. The coin has attracted a lot of Masternodes and the current number is over 4000 of them. Running a Dash Masternode is quite easy. One is required to have the Dash Masternode Tool installed in the server to act as the Masternode. The required coins can be locked in the common hardware wallets such as Trezor and Ledger. The tool also allows one to participate in the voting of proposals. In addition, there are many Masternode monitoring tools that can be used with Dash to help with the management of the Masternode.

6.2.2 ALQO

ALQO is one of the arguably small but high-potential coins. The coin is currently valued at $0.21 and one must have 10000 coins to host a Masternode. The cost comes to $2100 per Masternode. The good thing about the coin is the expected revenue. The coin has a daily income of $4, an annual income of $1702 and an annual ROI of 82%. Alqo is in the business of cryptocurrency exchange and it aims to create military-grade security for its crypto exchange. Alqo is also aiming to offer services that enable the payment of goods and services to merchants using cryptocurrencies. Alqo says that it will venture into many other businesses as well. Therefore, ALQO has a good potential for growth.

6.2.3 MUE (Monetary Unit)

MUE is not relatively new to the cryptocurrency space since it has been around since 2014. MUE coins are very cheap since each goes for $0.07. The coin has one of the highest number of coin requirements to run a Masternode which is 500,000. At its current cost, the Masternode costs $35000 to host. However, the coin stacks well in terms of revenues with market leaders such as Dash. The daily income is estimated to be $20.67 and the annual income is $7535 which is an annual ROI of 21.49%.

7 Practical set up of a Masternode

This chapter will give a hands-on tutorial on how to set up a real life Masternode. The coin selected is SmartCoin due to the ease of creating a Masternode on its network. The process is however similar across most cryptocurrencies.

7.1 SmartCash Masternode

To set up a SmartCash Masternode, one needs:

- A VPS
- SmartCash controller Wallet – this will be the Wallet on the local computer
- Masternode Wallet – this will be the wallet running on the remote computer (VPS)

The process begins with the installation of the SmartCash controller wallet on a local computer. When the installation is complete, one has to go to Help>Debug Window. This will open a console in which one has to type smartnode genkey and press enter. When this is done, the console will give out the smartnode genkey for a SmartCash Masternode. The next part entails preparing the Masternode. One has to create a new receiving address. This is done via File>Receiving address and a screen will appear with "new" at the bottom of the screen. This is as shown in the screenshot below:

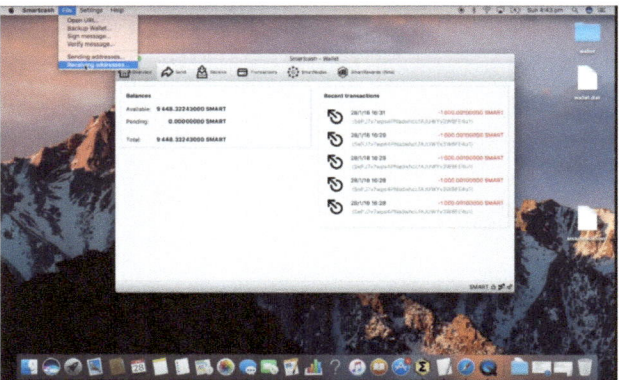

Figure 4: Creating a receiving address from the File menu

When a pop-up comes, one has to give the label "Masternode01" and click on OK. This will create the SmartNode's address and store it with the label Masternode01. This address should be copied as it is essential in the process and is the one that the collateral will be sent to. The screenshot below shows the pop-up to enter a label for the Masternode.

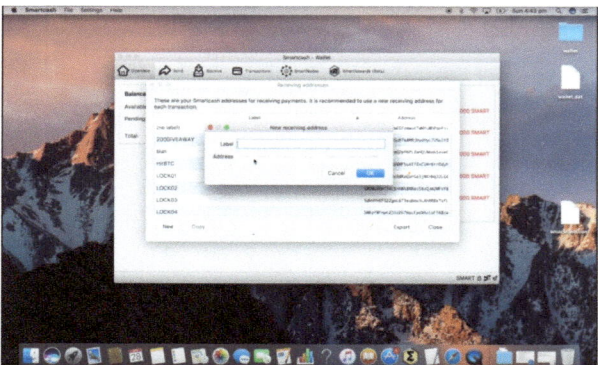

Figure 5: Pop-up requesting for the label

What follows is the transfer of the collateral to the Masternode01 address given. Therefore, one should send exactly 10,000 Smart coins to the Masternode01 address since that is the required collateral. Sending services can be accessed on the "send" tab and most importantly, not more or less than 10,000 coins should be sent to the Masternode01 address. After the sending is complete, on should go to the debug window so that the console can open. In the console, one should type smartnode outputs and save the values given in a Masternode.txt file.

The rest of the process entails setting up the actual server where the Masternode will run from. Smartnode works with Ubuntu VPS and this is available from vendors such as Vultr. One has to select a VPS with at least 1GB RAM and 20GB HDD. The best practice after renting the VPS is to update and upgrade it. This is for security reasons since updates and upgrades come with the latest security patches. These can be achieved by running the following commands on the Ubuntu terminal:

```
apt-get update
apt-get upgrade
```

The rest of the setup is to just get the Masternode running on the network. It is done by an automatic installer script which has to be downloaded from SmartCash's Github. The following command on the Ubuntu console will download the script:

```
wget https://raw.githubusercontent.com/SmartCash/smartnode/master/install.sh
```

The script can then be executed on the console using the following command:

```
bash ./install.sh
```

The install will run on its own but request the following inputs during the installation process:

- A custom SSH port (this is optional and can be ignored if not available)
- SmartNode GenKey (this was generated earlier on. It will link the local computer with the Smartnode on SmartCash)

The script downloads and installs multiple things and upon completion, it automatically reboots the server. When the server reboots, it is set up but not yet activated as a Masternode. Therefore, when the VPS starts, the SmartNode wallet will start and start syncing the Blockchain to get the necessary copies of files.

To finish up the process, one has to go back to the local computer and create the file smartcash.config in the folder %appdata%/Smartcash.

The following lines need to be added

Rpcuser=*TheUsernameoftheVPS*

RPcpassword=*ThepasswordoftheVPS*

The username and password can be found on the VPS server information from Vultr. In most cases the username is root and there is a password that one can copy directly from the vendor. This is as shown in the screenshot below:

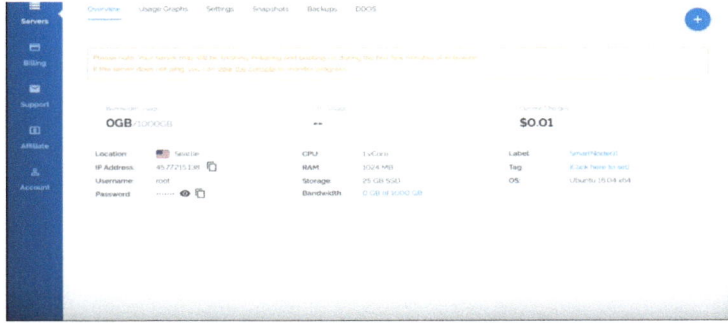

Figure 6: Screenshot from a Vultr VPS showing username and password location

Lastly, one has to open the smartnode.conf file in the SmartCash folder of the wallet and add the following information:

MasterNode01 xxx.xxx.xxx.xxx: xxxx *The IP address and port of VPS*

smartgenkey

output *The output of the smartgen key command on the wallet*

0   *The value of the "masternode output command"*

With this, the Masternode is all set up. All that remains is one to navigate to the Smartnodes tab on the wallet and start it.

8 Trends in Masternodes

8.1 Investing to own a Masternode

Since the cost of owning a Masternode is quite high, there is a new approach that investors are taking to reduce the costs for owning one. Instead of buying the Masternodes using their own cash, they are choosing to invest in cryptocurrencies to get the required amount or part of that amount. Unlike Bitcoin, there are coins that have been steadily gaining value at explosive rates. These are being used by crypto investors to grow their money to a point where they can afford a Masternode. There are Masternodes going for as little as $3000 and these are some of the targets of crypto investors. Another strategy being used is mining to get enough money to either own or to get close to owning a Masternode. However, mining is not as highly rewarding as before thus fewer people are choosing it as an investment option.

8.2 Speculation

Many of the reputable coins are charging too much for their collateral. This has turned many to look into the alternatives which are small coins that have growth potential. Therefore, there is a trend of speculative ownership of Masternodes in small coins. The owners of the Masternodes are speculating that with time, the value of the coins will go high up and as a result, they will benefit. The speculative holding of Masternodes has inadvertently led to more coins being established and offering Masternode investments. Therefore, careful analysis is required before one chooses one of the small and new coins to invest in. this is because there is a high risk that some of these coins will never pick and as a result, all those that will have invested in the coin will only run into losses.

8.3 Multiplicity of Masternodes

The coins that pioneered the use of Masternodes led to the belief that it is expensive to own a Masternode. However, entrants into this type of technology have made it feasible to own a Masternode with as little as $2000. To diversify investments, there is an observable trend of people owning more than one Masternode. Therefore, instead of pooling resources and hosting a Dash Masternode, there is a trend of people choosing smaller but promising coins and investing Masternodes in them. By doing this, they diversify their incomes and just in case one of these coins shoots up, they stand to benefit. There are no restrictions as to how many Masternodes one can own and this has made it easier for those willing to put a stake in different low-cost coins.

8.4 Adoption by other cryptocurrencies

Coins that did not previously have Masternodes are adopting them due to the benefits that they offer. These are setting up a mixed Blockchain technology comprising of both POW and POS. With the introduction of Masternodes to these coins, they are now able to offer services such as instant transfers that were not previously possible. It is predictable that in the long term all the coins will jump onto the Masternode bandwagon. This is because the costs of mining are going up and it will eventually not be feasible for miners to keep operating. The average transaction times are also high in coins that rely solely on miners while they are instantaneous on coins that have Masternodes. The predictions that have been set out already show that mining a single Bitcoin in 2020 will consume as much electricity as Denmark does in a day. There are many other negative predictions that make it unfeasible for any coin to keep miners on their network. Therefore, the future is guaranteed to be filled with Masternodes.

8.5 Rising cost of ownership

The initial cost to own a Masternode in reputable coins such as Dash is fast getting out of hand for many people. This is due to the big portion of the population being risk averse thus trying to buy into the coins that are more reputable. In the end, they are making the number of Masternodes in the networks of reputable coins to go high quickly. When the number of Masternodes is high, the revenues slump down since they have to be divided evenly across all the Masternodes in the network. At the same time, the coins grow in value as a result of more blocks being processed. The rise in the value of the coins is a relief to the people that already own Masternodes in the coin but is a deterrent to new entrants. This is because the cost of collateral will go up with the price of the coin. For instance, when Dash was new, one could own a Masternode with as little as $2000 when Dash was at $2 a coin. Many small coins will also gain in value and the cost of owning Masternodes in them will go up.

8.6 High ROI Masternodes scam

There are currently very many coins that have incorporated Masternodes. Many of these coins are new and unheard of. However, they are using a trick to lure people into opening Masternodes with them. These coins are offering mind-blowing returns of insane percentages such as 1500%. A close inspection shows that some of these coins are weak and do not have any tangible business plan. Therefore, people that invest in them with the hopes of getting these supernormal profits might end up getting burned when the coins destabilize and quickly lose value. This is one of the deterrents to the Masternode investment where people fear that after spending so much to buy the coins, the coins might lose value leading to them losing a lot of money altogether. Some of the coins offering these returns are also security concerns and could be easily hacked.

8.7 Artificial scarcity

The number of coins being demanded as collateral to own a Masternode for many cryptocurrencies is high. For instance, one needs 500,000 MUE coins to set up a MUE Masternode. The total coin supply for this cryptocurrency is 135 million. The number of enabled Masternodes on MUE is 125. Therefore, a total of 62.5 million coins have already been frozen. With an increase in the number of Masternodes due to the good ROI, the number of available coins will go down. In the end, there will be a scarcity of this coin and this will push its price up. The scarcity will have been caused by many coins being frozen by the Masternodes. This is something that never used to exist. Therefore, within a few years, there will be reports of coins that are scarce due to many people locking the coins in their Masternodes. The demand pressure for the coins in little supply will lead to the coins' prices going high up. This will be good for the Masternode owners but bad for the general populace.

8.8 Weak coins

All coins have a specific purpose for which they were created. They are then expected to be used primarily for that purpose. For instance, the APR coin was created for loyalty programs and the coin will allow for the transferability of loyalty points. Therefore, if a consumer that does not have loyalty points will want to acquire them, he or she will buy them using the APR coin.

This is a simple illustration of purpose and use of a coin. Another example is Bitcoin which was created specifically to be a currency. It is used as a currency in many places such as restaurants and the dark web.

However, with the advent of Masternodes, there has been a creation of coins that rarely set out to solve a certain problem that has not been solved by others in the real world. These coins are therefore weak since they have a purpose but will hardly see any use since the problem they wish to address already has been solved. Therefore, the cryptocurrencies are shielding these coins from this harsh reality using Masternodes. This is because the coins have no future in the real world and might not ever get to be used.

Masternodes, however, provide a way for these coins to be manipulated to gain value. With adverts of high ROIs, there will be many investors that want to buy into the coin to enjoy the advertised revenue. Once the number of Masternodes will start rising, the supply of the coin will reduce and the value will go up. However, this does not change the fact that the coin is still weak and will never get any real-world use. Therefore, if only a few people set up Masternodes with the coin and the supply remains to be excess, the coin will only stumble down. Therefore, caution must be exercised when one is buying these new coins. Particularly, one should be interested in knowing the problem that these coins aim to solve. It is also recommended that these new coins diversify their utilities. For instance, instead of APR only relying on loyalty programs, it could seek another related space where it can be used as well. This is also an opportunity for the coins that are weak and solve no real problems to find one that they can solve. Simply put, these coins need to find utility outside of trading. That way, their value can grow despite the happenings in the trading world.

Conclusion

This book has given an in-depth explanation of what a Masternode is. It has elaborated the types of Blockchain in existence and explained the type used by Masternodes. To give more context, the book has also explained the relationship between Full nodes and Masternodes. The importance of Masternodes in cryptocurrencies have been stated and explained. These are; increased privacy, instant transactions, more private transactions and decentralized governance of a coin. These are the roles that Masternodes play in cryptocurrencies and coins consequently lack these features. The requirements of owning a Masternodes have been explained at some depth by the book. The explanation given is based on the current environment and these requirements are similar across all cryptocurrencies.

An important aspect of Masternodes is their revenue model which is different from how miners get paid. Masternodes are assured to be paid and there is a queuing algorithm that assures of this. The mechanism of payment has been explained in detail. Masternodes require a great deal of care to run. This is because of their importance in the Blockchain of a coin and must therefore remain reliable. The best computing environment to run a Masternode is on a Virtual Private Server (VPS). The book has gone through VPSs, stated their pros and cons and then explained how one should choose the best VPS host. Masternodes are capital-intensive to own and that is why owners should strive to recoup their investments as quickly as possible and then start making profits. Not all coins are ideal for investing and this book explains how one should choose the best coin for a Masternode investment. In addition, the book gives three high-potential coins that investors should consider.

In the last chapters of the book, there has been an explanation of an almost-universal process of setting up of a Masternode on SmartCash. Lastly, the book has looked at the trends being witnessed in Masternode investments such as scam coins, ownership of multiple Masternodes, increased adoption of Masternodes by other coins and the looming threat of artificial scarcity. The in-depth discussions in the book are ideal for any willing Masternode investor and they cover all aspects of this type of business.

Acknowledgement
- Zizen
- Jo-Yu Duh
- Wei-Shiun Chen